UpClose #2

TIN LIDS

Canadian Combat Helmets

An "UpClose" publication of;

Service Publications
Nepean, Ontario

Introduction

I have been collecting combat helmets for almost three decades, and have never ceased to be amazed by the endless variety in shapes, markings, materials and liner and chin-strap arrangements which have been used over the past 80 years. There are a number of considerations that a helmet designer must take into account; the helmet must provide adequate protection without being unduly heavy and uncomfortable. It must protect the vital parts of the head without interfering with vision, hearing or the soldier's movements. The liner and chin-strap must be comfortable and keep the helmet securely on the soldiers head, yet still be robust and soldier proof. Other considerations include compatibility with respirators, head-phones, weapons-sights, providing a design which is readily distinguishable from that used by the enemy, and ensuring that the helmet's design is compatible with existing fabrication technology. Manganese steel, long the material of choice for combat helmets, is not easily worked and requires special presses and metal working techniques. All these considerations come up at one time or another in the course of this booklet.

Why write about *Canadian* helmets? Most people assume Canada has simply used other peoples helmets. On the whole yes, but even where Canada has used an existing designs, there have always been attempts to "Canadianize" it, and there have been at least three projects to develop a national helmet, plus various trials of tank and motorcycle helmets. Even the choice of other countries' designs shows a lot about the evolution of Canada's military procurement policies. I also hope this book will fill a gap for collectors. I started my research when I discovered how little reliable information was available on Canadian helmets. There has been no published information available (up to now) on which firms manufactured helmets and their components in Canada, or how many were produced. Even so, this study will not be definitive, there are still gaps in my knowledge which one day I hope to fill. I welcome comments from readers in this regard.

I would like to thank those who made this booklet possible: Clive M. Law of Service Publications, my publisher, for the inspiration to write this book and his invaluable guidance on how to navigate the holdings of the National Archives and DND's Directorate of History & Heritage; Ed Storey, CD of Kingston and Petawawa for his detailed knowledge of helmet use (and abuse) in the Canadian Army and the many archival and artifact photos that he made available; Marcus Cotton of London, England, for sharing his encyclopaedic knowledge on British helmets past and present; to fellow collectors and historians David Powers, of Laguna Beach, California and to Ross Wilson, Ken Joyce, Gary Boegel and Julie Tremblay of Ottawa for sharing information and photographs; the ever helpful staff of the National Archives and the Directorate of History and Heritage for helping me navigate the files; Margaret Mech, Ian Craigie and Brian Barnwell of DND/Directorate of Soldier Systems, Project Management for answering my many questions on modern helmet developments; and Harry Martin and Carol Reid of the Canadian War Museum for guidance on their holdings; to Colin Rowe, my photographer, for his excellent photographs which fill these pages, and; finally my thanks to my wife Susan and our cats for their patience, encouragement and, not least, their willingness to share our basement with 400 helmets.

Roger V. Lucy
Ottawa, Ontario
April, 1997

The Canadian Expeditionary Force 1914-19

For most of military history, one of the soldier's most basic pieces of defensive equipment has been his helmet. The wearing of protective helmets largely fell out of use in the 17th century, after advances in firearms made bullet-proof armour too heavy for any except specialist troops, such as heavy cavalry or sappers. This changed with the onset of trench warfare in World War I, and with it the steep rise in the number of head wounds caused by shrapnel and shell fragments. This created a new interest in providing soldiers with defensive equipment and, by 1916, all major combatants had adopted stamped steel trench helmets.

Figure 1 *The use of sandbags as a camouflage cover was quite common in the trenches.*
NAC PA356

The first generation of modern combat helmets which appeared in 1915, comprised three basic types: the French *Mle-15 Adrian*, the British *Mk. I, Brodie*, and the German *M-16 Stahlhelm*. Weighing respectively between 700 and 1400 grams they were not meant to be bullet proof, but did save many from death or serious injury from shrapnel and shell fragments. Each country adopted its own distinctive form and liner arrangement, reflecting the manufacturing capabilities, protective priorities and to some extent national aesthetic of the originator. As the Canadian Expeditionary Force shared the logistical train of the British Army, helmets worn by Canadians in the Great War, were, for all intents and purposes, the same as those worn by the British.

Helmet, Steel, Mk. I

Like other Empire forces on the Western Front, the Canadian Corps received its first issue of steel helmets towards the end of 1915. At first, supplies were very limited and helmets were considered part of each Brigade's "trench stores", to be turned over to replacement units, as battalions were rotated into and out of the line. As early as 24 November, 1915, Canadian Corps Routine Orders envisaged that, as supplies became available, each soldier would be issued his own helmet. As of 2 February, 1916, the divisions of the Canadian Corps still had only enough helmets to equip 16%-18% of their personnel. On 22 May, 1916 Canadian Corps Orders allowed officers to acquire their own private purchase helmets, but these were to conform, as closely as possible, to the regulation pattern. The GOC expressed the wish that they not buy the French pattern.[1]
The issue helmet, designated *Brodie's Steel Helmet, War Office Pattern*, had a shallow, wide brimmed shell, based on a design submitted to the British War Office by inventor John Leopold Brodie in the summer of 1915. With the exception of the first 4,440 produced in mild steel, it was made of non-magnetic manganese steel. Easily stamped with the pressing technology of

the time, it went into series production in October, 1915. The helmets can be approximately dated by a number of different changes that took place between 1915 and 1918. The earliest helmets were rimless and had a padded interior and the liner was made of oil-cloth, (America Cloth) with six tongues making up the crown. The liner was attached by a strap and rivet to the top of the helmet shell. The chin-strap comprised 2 pieces of leather, fastened with a claw buckle. A number of changes were approved in April, 1916. These included fitting a mild (magnetic) steel rim and larger lugs and wire loops to the shell. A new liner was designed, comprising a top pad of asbestos, lint and felt, attached by a copper rivet to the interior of the shell along with a leather strap, which held a headband of America Cloth, sewn to a basil (reddish brown) leather support, padded with cotton wool and covered by lint and canvas. There were no tongues, instead, an adjustable cotton net adjusted the band to the wearer's head. In the head-band support were set 12 tubular rubber buffers. The leather strap, which attached the liner to the shell, terminated in 2 small brass loops through which was passed and rivetted at the ends, a new one piece leather chin strap with a plated steel slide buckle. The linked straps passed through the wire loops attached to the chin-strap lugs. In September, 1916, this modified helmet received the designation *Helmet, Steel, Mk. I.* These changes were introduced between April and September, and

Figure 2 *Side-view of a typical Mk. I helmet, this particular specimen is painted blue-grey over its factory coat of khaki brown, note the heavy plated steel slide buckle on the chin-strap.*

Lucy Collection / Rowe Photo

Figure 3 *The liner is a post October 1917 with the rubber ring and cartridge cloth liner band, the cord adjusts a net sewn inside the crown. Note the split rivets holding the chin-strap lugs, and the letters FS 117 indicating that this helmet was made in the UK, with steel supplied by Thomas Firth & Sons Ltd. of Sheffield.*

Lucy Collection / Rowe Photo

Figure 4 *The front of another Mk. I this one is painted in sand textured greenish khaki. The shell is stamped HS 274 indicating that the steel came from Hadfields Steel of Sheffield. The white front insignia is the battle flash of 5 Division artillery, the only part of that formation that got overseas.*
Lucy Collection / Rowe Photo

combinations of old and new features can be found on helmets manufactured in this period.

Helmet linings made prior to August 1916 have the words PATENT APPLIED FOR, later liners are marked PATENT NO. 11803/16. Further modifications were the addition of a grey rubber ring inside the top pad from June 1917 and, at about the same time, the substitution of an artificial leather called "cordite cloth" for basil leather in the head-band support.[2] The helmet bodies were marked by large letters and digits stamped on the brim to indicate the steel maker and lot number, such as FS for Thomas Firth and Sons or HS for Hadfields Ltd (both of Sheffield).

Collectors should note that the US-made version of the Mk. I the M-1917 can appear in Canadian contexts. US helmets can be distinguished from British-made helmets by their serial number which has a 2 or 3 letter code (with or without numbers) beginning with U, X, Y or Z. Where the original fittings exist, the US chin-strap lug is heavier and held by a double ended machine rivet, British-made helmets use a split-tailed rivet.[3]

As with their British counterparts, many early Canadian helmets were painted a blue-grey or dark green colour over their khaki factory finish. From 31 October, 1916, sand textured khaki-brown (often over-painted in other colours) was recommended. Frequently, regimental or arm of service badges were painted or stencilled (or even - in the form of the cap badge - welded) to the front. From 1918 the Canadian divisional and unit battle flash, worn on the tunic sleeve since mid-1916, was painted on the helmet. (Service Publications' *Distinguishing Patches* provides a guide to identify these symbols.) At the same time efforts were being made to camouflage the helmets. Sand-bag covers (made at the unit level or by Army Ordnance Depots) were adopted in late March, 1916. The Canadian Corps also conducted trials with the Cruise chain visor in March, 1918. The visors were found to offer insufficient protection, obscured night vision, were too fragile for operational use, and were incompatible with the use of rifle sights and gas masks. Adoption was not recommended.[4]

With the end of the Great War, the Mk. I helmet soldiered on in the ranks of the much reduced Canadian Army. Pictorial evidence shows that these could be adorned with arm of service or regimental badges. By the outbreak of World War II only 25,000 helmets remained in National Defence's depots. To meet the demands of the rapidly expanding Canadian Army new helmet production was required.

Helmet, Steel Mk. II

Figure 5 *These helmets display the hand painted camouflage scheme that was common in England during the early days of the war.*

NAC PA197526

Even before the outbreak of World War II, it was becoming increasingly evident that, in the event of mobilisation, Britain could not meet all of Canada's equipment needs. Many basic items of arms and equipment, whose WWI equivalent had been provided from British stocks, would have to be produced in Canada. After June, 1940, when the British Expeditionary Force had to abandon most of its stores at Dunkirk, the supply situation became even more acute. Initially however, Canada could not fully equip (at least not with modern equipment) its 1st and 2nd Divisions, sent overseas in the winter of 1939 and the summer of 1940. Some troops went overseas in Mk. I helmets, while 140,000 new British-made steel helmets were purchased. These helmets were of the revised *Helmet, Steel, Mk. II* pattern which had a non-magnetic stainless steel rim and lugs, as well as a new liner of fibre and oil cloth and a sprung canvas chin strap. The Mk. II was introduced in 1938 and underwent a number of modifications over time.[5] Specimens and specifications of the new British Mk. II helmet were obtained in 1939 and forwarded to Canada. The National Research Council conducted tests to determine the correct grade of non-magnetic steel.

Figure 6 *A Classic example of a MK. II with a Regimental decal affixed to the side. In this case the Victoria Rifles of Canada.*

Lucy Collection / Rowe Photo

Formal procurement began on January 25, 1940, when the Assistant Deputy Minister (ADM) Militia in the Department of

National Defence issued a requirement for Canadian manufactured Mk. II steel helmets, to be delivered to the National Defence Inspection Room in Ottawa for issue to the Canadian Active Service Force. The Department of Munitions and Supply (DMS) issued a call for bids for the manufacture of 25,000 non-magnetic steel helmet bodies (with lugs and securing bolts), and 51,250 liners and the chin straps (the extras were for retrofitting existing Mk. I helmets). The winning bids were approved by the Privy Council on 25 May, 1940. Canadian Motorlamp Co. of Windsor Ontario (with a bid of $24,300 - including taxes etc.) got the contract for the bodies. Viceroy Manufacturing Co. of Toronto (with a bid of $32,678.03) won the contract for the liners while Backstay Standard of Windsor, Ontario, (with a bid of $21,252) won the contract for the chin-straps.[6]

Figure 7 *Interior of a typical Canadian Mk. II note the shape of the chin-strap lugs.*

Lucy Collection / Rowe Photo

National Defence issued a follow-on requirement for 115,000 helmets and 120,000 liners and chin-straps in late June. Viceroy Manufacturing Co. and Backstay Standard again won (as they were to do throughout the war) the contracts for the liners and chin-straps. Their bids were consistently a good 25-30% lower than those of their nearest competitor (Scully's of Montreal). On this occasion, the lowest bid for the bodies - at just under $0.92 each - was made by General Steelwares of Toronto. However, in light of Canadian Motorlamp Co.'s past experience in making helmet bodies, DMS recommended that the order be split between the two companies, with the Canadian Motorlamp Co. lowering its price to meet that of General Steelwares and undertaking to improve its delivery time. Each company therefore received a contract, worth $52,756.25, to produce 57,500 helmet shells. The contracts were approved by Order-in-Council in late August.

Figure 8 *A Viceroy Manufacturing Company liner, note the acceptance stamp.*

Lucy Collection / Rowe Photo

Production tempo picked up, at the end of 1940. Another 136,000 helmets, chin-straps and liners were ordered, with 20,000 destined for the RCAF. The contracts for the bodies were again equally divided between General Steelwares and the Canadian Motorlamp Co. In 1941 a total of 200,000 helmet bodies, 250,000 chin-straps and 350,000 liners were ordered. Both Viceroy and Backstay continued to be sole-sourced for the liners and chin-straps. Only General Steelwares won contracts for the helmets, as they had a more secure supply of manganese steel (which had to be imported from the USA) and could offer better guarantees of delivery. A minor change was instituted in liners ordered after July, 1941, these were fitted with steel wire clips to retain the rubber buffers on the liner band - a feature never found in UK-made helmets.

In 1942, the cost of the helmet bodies rose some 30%. Supplies of manganese steel supplies had become scarcer because of the USA's entry into the war, and the supplies of steel that were available had to be subjected to a pickling process. Altogether this raised the cost of helmet bodies from $0.95 to about $1.30 each. The Canadian Motorlamp Co., won almost all the contracts for the 500,000 helmet bodies ordered by DMS in 1942. This was due to superior quality control (the Department of Munitions and Supplies reported they had a rejection rate of less than 5%), better access to steel supplies and better delivery guarantees. Of these, 300,000 were ordered by the Army, and 200,000 for the RCAF. The Navy presumably found its helmets the best it could. The Department of Munitions and Supply did try to expand the number of shell producers, encouraging bids from other metal firms. Only the Aluminium Goods Ltd. of Toronto managed to win a tender in May, 1942 for 66,600

Figure 9 - 11 *The logos of the 3 Canadian helmet body manufactures - Aluminum Goods Company, Canadian Lamp Company and General Steelwares. The "C" on the AG/C and the "78" on the CL/C helmets are the batch identifiers. CL/C numbered their helmets consecutively from 1940-43. GSW which also used a letter code reused its letters each year. The letters DP&H represent Department of Pensions and Health, these were the mild steel stampings issued for ARP purposes.*

Lucy Collection / Rowe Photo

Figure 12 *A Machine Gun crew of the Royal Montreal Regiment training in the UK. The Machine Gunner is wearing the RMR decal over the unit's battle flash.* *NAC PA197523*

helmet bodies for the Air Force. While their price was very favourable this was their sole helmet contract, their shells are marked by the logo A.G./C. From 1942 General Steelwares produced only Civil Defence helmets (see below).

The last orders for helmet bodies was placed in September 1942, and the last 90,000 shells were completed early the following year. Thereafter, aside from 125,000 replacement liners and chin-straps ordered up to 1945 (produced at the rate of about 8-9,000 chin-straps and about 10,000 liners per month), and another batch of liners and spare parts ordered in 1953, Mk. II helmet production ended in Canada. In all, 976,600 Mk. II helmet bodies were ordered for the military, and 155,000 mild steel variants for Civil Defence purposes - comprising a total of 1,131,600 bodies. Including those for Civil Defence, 1,400,240 liners and 1,275,785 chin-straps were ordered during the war. The official Canadian designation was *Helmet, Steel, Mk. II* with a stores number *IA8350*. The bodies being designated: *Body Mk. I. to Helmet, Steel, Mk. II*; the liner: *Liner Mk. I* or *Mk. II.* There is no indication of how they differed unless the Mk. I

Figure 13 *Sand-bag covers redux. Newfoundland, circa 1941. This form of covering fell out of favour as it was seen a posing a potential decontamination problem.* *NAC PA197525*

Figure 14 *A Canadian-made Mk. II (Aluminum Goods Company) painted in a thick textured sand colour, with additional camouflage provided by a British-made No. 1 anti-gas curtain. Period photographs show covers like this worn on helmets at the start of the Sicilian campaign.*

Lucy Collection / Rowe Photo

is the version without the buffer retention springs. The chin-straps were marked; *Strap, Chin, Mk. II.*[7] Helmet bodies were marked with the manufacturer's logo, their year of manufacture and a batch number or letter. Canadian Motorlamp Co., helmet bodies bore the marking C.L./C. and a number, General Steel Wares and Aluminum Goods used respectively the logos G.S.W. and A.G./C. and a letter. The liners bore the logo V.M.C., the date, the size (63/4 to 73/4) and often a white C/Arrow, usually with an inspector's number. Chin-straps are usually marked only with the C/Arrow.

British-made Mk. II helmets have different markings (for example B.M.B for Briggs Motor Bodies or RO & CO for Rubery Owen and Co). To meet war-time supply contingencies they progressively underwent a number of changes to the lugs, chin-strap and securing bolt. Canadian produced helmets retained, throughout their 1940-43 production run, the features of the British pattern, as it was in late 1939, with clip-cornered square Mk. II lugs, the web and springs Mk. II chin-strap and the wide brass Mk. I lining securing bolt with domed nut. The Canadian-made Mk. II helmet, was made from Hadfields manganese steel 0.033" (0.825 mm) thick. With chin-strap and liner it weighed about 1,100 grams (approx. 2¼ lbs). According to the Department of Munitions and Supply's official report, the Canadian Mk. II was supposed to resist penetration from a .303 bullet at 12 feet. This seems improbable. Tests carried out in England in October 1942 showed it could resist penetration from a pistol bullet at 1 yard and from a

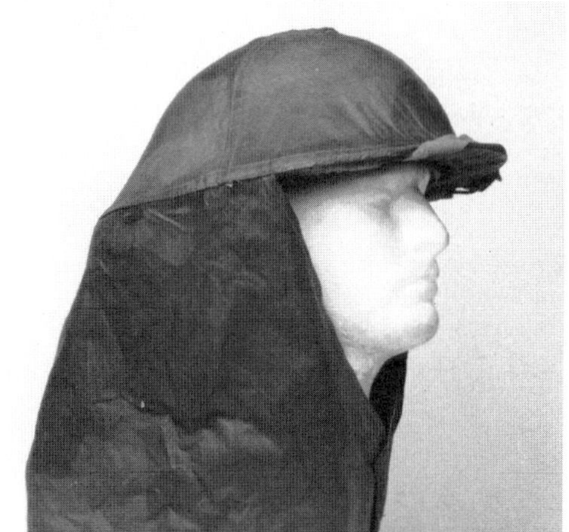

Figure 15 *A Canadian made Mk.II (General Steelwares, 1940) covered by a brown and green patterned No.2 anti-gas curtain, made by Aristo of Galt, Ontario from 1941, others were made by the Scythes Company, Limited of Toronto and Montreal and Dominion Rubber Co., in both camouflage and plain olive drab.*

Lucy Collection / Rowe Photo

Thompson SMG at 3 yards. Ballistic tests, carried out at the US Army arsenal at Watertown in April, 1952, using a .22 calibre 17 grain fragment simulating projectile and 17 grain US and Soviet shell fragments showed the WWII Canadian Mk. II had a v50* of about 900 ft/sec (270 m/s). By comparison a US M-1 helmet had a v50 of 350 m/s and today's new generation of ballistic composite helmets have v50s of up to 700 m/s. A Mk. II variant that does not seem to have reached production is illustrated in the 10 December, 1942 "Hamilton Spectator". Reportedly made of a "new type steel", it is shown, relatively unscathed from a burst of Tommy-gun fire (the range is not specified), beside a somewhat perforated regular Mk. II. No other details are known about this.[8]

Figure 16 *An early Canadian-made Mk. II, marked C.L./C. 1940. The body has been over-painted in sand textured dark olive green, with a blue striped crimson Royal Canadian Engineers decal on the left.*

Lucy Collection / Rowe Photo

Helmets were needed, not only to meet mobilization needs, but to cover wastage. Soldiers normally went overseas with their own helmets, but even in a non-combat situation, the brutal and licentious soldiery had a proclivity to subject their helmets to unfair wear and tear (for example in 1916, Canadian Corps passed on a BEF order forbidding soldiers to heat their helmets over open fires). In August, 1941, Canadian Military Headquarters (CMHQ) in London reckoned that for the next 6 months it would need 2,500 helmets, 5,000 liners, 15,000 lugs and rivets and 8,000 securing bolts to cover normal wastage. By August, 1944, however, the helmet supply situation was so favourable that 25,000 helmet bodies, liner and chin-straps were passed by the Air Force to the War Assets Corporation for disposal.[9]

While the Department of Munitions and Supply was responsible for helmets required by the Armed Forces, the Department of Pensions, Health and Welfare was responsible for supplying helmets for the Air Raid Protection Service. Initially it was supplied with helmets from National Defence stocks, receiving 15,600 in 1941. After Pearl Harbour, however, the area potentially threatened by air raids greatly increased, while shortages of specialty steels forced up the price of military-issue helmets. Pensions and Health therefore obtained authorization in January 1942 to procure 30,000 helmet bodies, rejected by National Defence, from General Steelwares. To achieve further economies, it tendered General Steelwares for helmets made from hot rolled

*v50 is a common unit for the measure of protection offered by a helmet or body armour it represents the velocity at which the helmet will stop half of all 17 grain steel - or 9 mm bullet - fragments fired at it.

steel, which it could manufacture for $.50 each. This steel was magnetic and lacked the smooth finish found in the steel used in military helmets, but sufficed for Air Raid Protection purposes. Visually these helmets can be distinguished by their chin-strap lugs being spot-welded, rather than rivetted, to the brim. They are also marked G.S.W. D.P.& H. and lack the date or batch number found on military helmets. In all, in 1942, Pensions and Health ordered a total of 155,000 helmets from General Steel Wares, and equivalent supplies of liners and chin straps from Viceroy and Backstay Standard respectively. The last orders were placed in November, 1942, with final deliveries in 1943.

Figure 17 *England early war, this sargeant of 1st Field Regiment RCA has the unit tactical sign painted on his helmet, a white 5 over the red over blue RCA colours.*

NAC PA197528

Canadian helmets (including those produced for Air Force, and Air Raid Protection contracts) received a base coat of sprayed, smooth matte khaki. The earlier production helmets are a browner shade of khaki, while those produced after 1942 tend more to a light olive drab. Helmets usually retained their smooth khaki finish, but Navy helmets may be found repainted blue and police helmets black. Camouflage colours are occasionally encountered or a repaint in sand-textured olive-green or infrequently tan. Regimental or arm of service decals were frequently worn, usually on the left side. Canadian Army (Overseas) Routine Orders allowed the application of regimental decals or other flashes to the helmet. Soldiers usually painted their name and serial number inside the rim or inked it on the liner or shell interior. Very rarely the unit designation was stamped inside the liner.

A camouflage net was frequently worn. The earlier patterns were made of close-meshed woven or knotted khaki or olive green cotton. By 1943 a bi-coloured brown and green net (usually knotted but more infrequently woven) was in almost universal use. To this was often tied scraps of cloth scrim, while a field dressing was frequently worn beneath it. Wide meshed knotted nets were sometimes used in NW Europe. Another helmet cover was the impregnated oil cloth anti-gas curtain issued to protect the wearer from mustard gas and other vesicants. The *curtain, anti-gas, No.1* just covered the rim of the helmet with a curtain protecting the sides and back. It was available only in olive. The No.2 anti-gas curtain, which came in both plain olive and in a green/brown blotched camouflage version covered the entire helmet. This cover was produced in Canada from 1941. These curtains were sometimes worn just as camouflage covers, (especially by naval personnel to protect their helmets from salt water.)

M-1 or Mk. III?

Figure 18 *PR shots, early 1943, of First Canadian Army soldier modelling the new American, taken in anticipation of the announcement that Canada would be adopting this helmet.*

NAC PA197529

While not a bad trench helmet, giving good protection against falling shrapnel, the Mk. II gave little protection from the horizontal missiles more often encountered in mobile warfare. After studying the problem, the British Medical Research Council (MRC) came up with a new helmet design which provided better protection to the back and sides of the head. An initial run of this *Helmet, Steel, Mk. III* was issued for user trials and evaluation in the latter half of 1941. However, with the USA's entry into the war, the British Army began to study the feasibility of adopting the new American *M-1* helmet. In October, 1942 the War Office informed CMHQ and other Dominion HQs in London of the Army Council's decision to adopt the US M-1. The British were not entirely happy with the design but were prepared to accept the M-1 as-is at first, until they could undertake their own manufacture with required modifications.[10] As it was Canadian policy to conform to UK equipment in the European theatre, the adoption of the M-1 was almost axiomatic. National Defence Headquarters (NDHQ) in Ottawa was advised of the British decision on October 6 and was sent a copy of the British evaluation of the M-1 on 26 October. By 14 October, 1942, First Canadian Army ordered 1001 M-1 helmets, complete with liner and chin-strap, from US Army HQ Services of Supplies, US European Theatre of Operations. These were to be delivered by October 30, to No.2 Sub-depot Brookwood, Surrey. These helmets, worth $4,845 were distributed for troop trials on 21 October, 1942. 700 went to 1 Canadian Corps and 150 each to 4th and 5th Canadian Armoured Divisions. First Canadian Army asked recipient units to report on; the degree of protection afforded; advantages and disadvantages of the Mk. II; recommendations for adoption by the Canadian Army, and; proposed modifications to the M-1.

First Canadian Army's report on these trials, submitted to CMHQ on 30 November, strongly endorsed adoption of the M-1. Compared to the Mk. II, the M-1 was more comfortable to wear, provided better balance and ventilation, was more easily adjusted and offered better protection to the ears against blast. Compared to the Mk. II, it could be camouflaged more easily, and movement through undergrowth was less difficult. Moreover the outer shell could be used as a bucket and inner shell as a sun-helmet. Ballistic tests showed the M-1 and the Mk. II were more or less comparable. Both resisted close range .38 and .45 calibre pistol shots, but the Mk. II suffered more penetrations when a grenade was detonated one foot away.

On the minus side, the M-1 was judged; more awkward to carry; seemed more vulnerable to a

Figure 19 *A British-made Mk. III painted in sand-textured dark charcoal grey, painted over khaki. The painted red/blue insignia with a black outlined 43 is the unit serial of the 13th Field Rgt. RCA, an SP equipped unit of 3 Cdn Division.*
Lucy Collection / Rowe Photo

direct hit; inclined to echo; restricted vision; it required took two seconds more to don a British-type respirator; it tended to shift forward when firing in a prone position; looked German (a common complaint about any helmet which was not a Mk. I or II); could not be worn with standard British "watch-case" head-sets, and; the non-elastic chin-strap was uncomfortable and potentially dangerous (making it more liable to cause concussion by throwing back the wearer's head in a blast). Further its lack of a wide brim allowed water to drip down the wearers neck. While the overall conclusion was that the Canadian Army should adopt the M-1, several changes were recommended; install a top pad and an elastic chin-strap; use Canadian grade steel, and; deepen the rear brim to divert rainwater. It was also recommended that some of the helmets be adapted for use with British style head-phones, and others be provided with a reinforced harness for use by motorcyclists. On the basis of this report, CMHQ Staff Duties recommended to National Defence Headquarters, on 7 December, 1942, that the M-1 be acquired as-is to meet immediate requirements, while the suggested modifications would be made in due course. On the same day, CMHQ's Military Equipment Policy Letter 19 advised Canadian units in the UK that the M-1 would replace the Mk. II, except for headphone users and Armoured Fighting Vehicle (AFV) crews.

On the basis of a tasking by CMHQ's Technical Department, procurement action began at National Defence Headquarters. The Army Technical Development Board initiated Project 1401 on 18 December, 1942 to develop a new Canadian steel helmet based on the M-1 pattern, but using Canadian steel; incorporating a top pad to reduce concussion; reinforcing the harness for use by motorcyclists; redesigning the helmet to permit the wearing of head-phones, and; providing a curved rear skirt to prevent drip. While no design for this helmet has emerged, it is hard to imagine it not having an even more unacceptably Germanic shape.[11]

Pending Canadian production, steps were taken for direct procurement of 250,000 un-modified M-1s from the US Government. National Defence Headquarters issued a financial encumbrance

Figure 20 *A classic shot of a 3 Canadian Division mortar crew in in April, 1944 wearing net-covered Mk. IIIs.*

NAC PA197534

of $935,730 for their purchase on 13 December, 1942. This was approved by OIC 1212 on 15 February, 1943. The Canadian Military Liaison Office in Washington was already lobbying for an allocation. The USA was hard pressed to meet this order given the urgency of equipping its own rapidly growing army, coupled with the need to fill an order of unknown magnitude (but probably in the order of 1 million helmets) for the British. On 22 January, 1943 the US Military Assistance Command did approve the assignment to Canada of 250,000 M-1s (with 100,000 from the January production and another 100,000 from February's) only to withdraw it four days later, pending resolution of the British order. CMHQ rejected the Washington Liaison Staff's suggestion that it try for an allocation of 50,000 per month, and instead tried to get the British War Office to give the USA a clear indication of their own M-1 needs or, failing that, to back Canada's request for its own allocation. The War Office's answer of 18 February to CMHQ's queries should have rung some alarm bells. The British were no longer planning to acquire helmets from the USA and that production orders were being placed instead with the British Ministry of Supply for new helmets. This exchange left CMHQ with the impression that the War Office was planning to adopt the M-1 when, in fact, the Army Council was having second thoughts about it. British trials had voiced criticisms of the M-1 similar to the Canadian ones and the War Office had recommended a number of changes to the design (again similar to those under consideration by the Army Technical Development Board) to allow the use of head-phones and eliminate post-helmet drip. The USA had categorically refused to make these changes. British production was considered, but the deep draw presses needed to stamp M-1 bodies were not available in the UK. In view of the USA refusal to modify the M-1 to meet British requirements, the Deputy Chief of the Imperial General Staff decided on 3 March, 1943 to adopt the Mk. III, the improved helmet design developed by the Medical Research Council in 1941, but shelved in the interests of standardization on the M-1 pattern. The British were quite scathing about USA claims of the superiority of the ear-plug type head-sets they had adopted. The British claimed to have information indicating that the M-1's designers had overlooked the need to accommodate head-sets, forcing them to adopt ear plugs, which, in their

opinion, were less efficient and less comfortable than the "watch-case type" used by the UK, and indeed potentially dangerous to the user.
Despite the War Office's impending change of heart, the news was slow to trickle into CMHQ and CMHQ was slower still to act on it. As of mid-February CMHQ was still being told at senior levels that the Imperial General Staff intended to adopt the M-1. CMHQ so advised Ottawa. On March 2, National Defence Headquarters advised CMHQ that the USA had agreed to allot 200,000 helmets to Canada immediately. The telegram from CMHQ's files has a note initialled by a lieutenant on March 4.

> *"Have learned indirectly that W.O. are NOT adopting American pattern steel helmets and are producing a Mk. III which is old British Helmet with a deeper crown"*

This was confirmed directly on March 6 when the War Office copied the Staff Duties (SD) of CMHQ on a cable it had sent the British Military Staff in Washington

> *"Decision now taken no, repeat no, British Army requirement American pattern steel helmets. Dependant on allocation manganese steel production of British Mark III helmet will be pressed in UK"*

A subsequent British cable asked that the UK allocation of M-1s be passed to Canada. Despite all this, as late as March 26, CMHQ Staff Duties GSO 1, Lt. Colonel Anderson, still seemed to believe (although he had initialled the March 6 War Office message) that the British were acquiring the M-1, but from British production. He sent a memo to CMHQ's Technical Section asking them to consider, in conjunction with National Defence Headquarters, what modifications should be made to the 250,000 M-1s that the Canadian contingent would soon be receiving, to accommodate headphone wearers. National Defence Headquarters was meanwhile asking the

Figure 21 *A US-made M-1 helmet issued to a Canadian officer stationed in Nanaimo, BC in 1943. The contract numbers on the liner webbing indicates that it was made in late 1942, and could be one of the 200,000 which landed on NDHQ's doorstep in April 1943. The helmet is painted a dark olive drab.*

Lucy Collection / Rowe Photo

liaison office in Washington what camouflage colour the US helmet was painted. It was a Major MacLeod, a signals officer in SD, who on 29 March, raised the concern that the M-1 could not easily be modified for head-phones, that the Canadian Corps of Signals shared UK concerns that US plug-type headsets were inferior to the British "watch-case" type. He pointed out (again) that the British were therefore opting for the Mk. III and that Canada should do the same. This time the penny dropped. A letter of 29 March from Lt.Col F.F. Fulton, GSO1 in the technical section, referred the matter to HQ First Canadian Army Headquarters outlining the situation and suggesting four options:

> *a) Request Canada to modify all American pattern helmets, prior to shipment to this country. This would involve tooling up and manufacture of dies , the stripping of the steel helmets, the pressing of bulges for the head receivers and possibly as well a flare at the rear of the helmet; possibly re-annealing and re-painting; cutting away the sides of the inner helmets and re-assembling. It is felt this process would require time and in effect expenditure of unproductive effort, and that, therefore, Canada would not be willing to undertake this work.*
>
> *b) Issue American pattern steel helmets to all personnel other than users of head sets and issue Signals and other head set users the British Mk. IIs or British Mk. III pattern. This is considered highly undesirable by reason of the distinctive appearance personnel utilising a wireless . . . making them a conspicuous and probably desirable target.*
>
> *c) Obtain and issue ear-plug type head sets for all requirements and issue American pattern steel helmets throughout.*
>
> *d) Retain the British pattern Mk. II in use with watch-case receivers and provide British Mk. III from UK production, when they became available (late 1944).*
>
> *(a) and (b) above appeal to us as both being unadvisable. A decision is therefore requested as to whether proposals (c) and (d) should be adopted or if you have any further suggestions to make.*

On 3 April, 1943 General McNaughton, GOC First Canadian Army, decided that:

> *"The ruling policy in Steel Helmets is that Cdn Army be uniform with British, unless another policy should give substantial advantage. There is no such advantage in the U S type and in consequence we will adhere to British Mk. II until Mk. III becomes available."*

That same day, CMHQ cabled National Defence Headquarters to arrange the cancellation of the order for 250,000 M-1s from the USA. The Mk. II would be retained, until stocks of the Mk. III became available, probably in early 1944.[12]

On April 5, National Defence Headquarters cabled back that cancellation was impossible - 100,000 M-1s had already arrived in Ottawa and were being prepared for overseas shipment, while the second allotment of 100,000 was due very shortly. The final 50,000 of the order could while the second allotment of 100,000 was due very shortly. The final 50,000 of the order could be cancelled. National Defence Headquarters planned to send the now unwanted 200,000 M-1 helmets to the UK. First Canadian Army would either have to use them or pass them on to USA ETO forces. National Defence Headquarters asked for a sample and the specifications of the Mk. III and asked if there was a requirement for its production in Canada. On 10 April CMHQ cabled National Defence Headquarters asking it not to ship the 200,000 M-1s (i.e. over 300 tonnes of materiel) but to try to return them to the USA. With almost $1 million worth of unwanted helmets on its hands, National Defence Headquarters retorted on 16 April that *"the whole transaction was most unfortunate"*. Lt.Col Anderson tried to imply to his superiors that it was all the fault of CMHQ's Technical Department, noting "*in future technical specifications should not be interpreted as "ORDER" on Canada*". On April 26, General McNaughton replied to Ottawa, claiming that the reversal was wholly due to the change in UK plans. No reference was made to the three weeks that it had taken CMHQ to realize that the British had done so. National Defence Headquarters was still very unhappy about the fiasco and on July 12 was still trying to obtain from CMHQ the details of how the M-1 procurement decision had been made and reversed. CMHQ decided the less said the better. Its reply to Ottawa, sent on July 27, stuck to the party line: CMHQ had opted for the M-1 and then changed its policy in favour of the Mk. III in order to remain in conformity with British practice.

Figure 22 *First Canadian Army soldiers modelling the new American M-1 next to the Mk. II.*

NAC PA197530

Meanwhile, on 30 April, 1943, CMHQ Staff Duties advised the Deputy Quartermaster General that there was no staff requirement to produce the Mk. III in Canada. CMHQ Equipment Policy Letter 55 of 3 June, 1943 (originally drafted on April 5) cancelled letter 19 of 7 December, 1942 and confirmed that the Mk. III would replace the Mk. II on a one to one basis as they became available from British stocks. Issues were in fact slow. Because the deep draw stamping equipment and special manganese steel needed for the helmet had to be imported into the UK, it only went into full production in November 1943. According to a 25 February, 1943 answer to a National Defence Headquarters query, CMHQ reported that issues had only just begun and were confined to 3 Division in accordance to 21 Army Group's priorities, giving precedence to

Figure 23 *Canadian soldiers, returning from Kiska 1943, wearing M-1 helmets and Canadian bi-coloured nets.*
Photo courtesy Ed Storey, 220-6

units destined to land with the first wave at Normandy. As the helmets remained property of the British War Office, most were returned to Britain at the end of hostilities. For the rest of the Canadian Army, the Mk. II remained the standard helmet until September 1960, when the M-1 officially replaced it.

Despite this limited issue, the turtle-shaped Mk. III steel helmet became so much a part of the image of the Canadian infantryman in Northwest Europe and so inextricably linked with the Canadian Army, that many collectors refer to it as the "Canadian" or "3 Div Helmet". Common collector mythology has it that the Mk. III was a Canadian design. Obviously this is not the case, and all Mk. IIIs worn by Canadians were produced in Britain by either Briggs Motor Bodies or by Rubery Owen. The Mk. III specifications and samples, which Ottawa had requested, were sent on 16 April, 1943. As noted above, National Defence agreed with CMHQ that, to conserve scarce shipping space, the helmets would be sourced in England and not manufactured in Canada.[13]

As to the 200,000 unwanted M-1s? Most appeared to have been used in Pacific Command. Some may have been passed back to the Americans for use by their troops stationed in Canada. In September 1944, the British Liaison Staff requested 30,000 surplus Canadian M-1s for issue to Wingate's Chindits in Burma. Others went to equip Canadian troops on the Pacific coast or elsewhere in Canada. Aside from 13 Brigade, which took part in the liberation of Kiska in the Aleutians in July 1943, (although it may have received its M-1s, along with its other US equipment, directly from the USA), other West Coast units, such as those of 18 Brigade, wore the M-1 both in Canada and in the Caribbean. In 1945, M-1s were also issued to the specially constituted 6 Canadian Division, which was to have been part of the Canadian Army Pacific Force. As of 18 October, 1951, there were still at least 20,000 M-1 helmets on hand in Canadian Army depots. Its official Canadian Army stores number was *IA18750*.[14]

Early WWII-vintage M-1s, of the type delivered in April, 1943, can be identified by the following features: inner fibre liner - khaki webbing; bare steel furniture; chin-strap and vent outer steel shell - khaki chin strap sewn to brackets welded to the sides of the shell; cast brass buckle, and; a rim butted at shell front. A lot of early M-1s found in Canadian contexts have the fibre Hawley liner or limited production "low-pressure" resinated liners, types that had been declared "limited standard" as insufficiently robust for service use. These lack the "wood-grain" pattern found on the interior of "high-pressure" liners. To meet Canada's urgent demands for M-1, the Americans may well have scraped the bottom of their depots.[15]

Motorcycle

***Figure 24** England early war, a DR corporal wearing the British Economy Pattern DR helmet made of rubber and fibre pulp.*
NAC PA197524

Prior to 1940, Canadian Despatch Riders (DRs) wore no special protective head-gear other than the standard steel helmet, when combat conditions so dictated. First Canadian Army soon noted that fibre crash helmets were being worn by British Army motorcyclists. On March 30, 1940 the Chief of Staff of First Canadian Army recommended that crash helmets be procured and that 700 would be required for the DRs and Military Police in each Canadian division. CMHQ accepted this recommendation. On April 17 it ordered 750 off-the-shelf helmets, "conforming largely to the British War Office pattern", from Charles Owen & Co. in London The helmets cost 17/9d each and were to be delivered at the rate of 100 per week. Concerns were raised that some DRs could be shot by mistake because the helmet's outline was too like that of the German para helmet. CMHQ dismissed these in July:

> *"...despatch riders will have to chance their heads, or else trust guards and sentries gradually becoming familiar with this rather foreign looking piece of equipment."*

In the meantime, Ottawa attempted to find a Canadian company willing to manufacture crash helmets, but, as of June, 1940, had had no success. An additional 750 helmets were ordered from Owen on June 19, 1940 to meet 2 Division and Corps troops requirements. These orders were completed by January, 1941. Further requirements were met from War Office stocks. A quote on 5,500 DR helmets was requested from the War Office in December, 1940 and 750 were ordered on January 31, 1941 at £1 8/ each. Issue of these was complete by March and another 750 were ordered in June. Even so, with the rapid expansion of the Canadian presence in Britain, it was estimated, at the end of August, 1941, that an additional 1,000 motorcycle helmets were required for the next six months.[16] In June, 1942, the War Office agreed that CMHQ could order 6,000 WO pattern motorcycle crash helmets (at the rate of 600 per month), from the firm Helmets Ltd., of Oxford, at 14/ each.

Supply of motorcycle helmets, however, remained a problem through to 1944. British made "pulp" DR helmets, have a high egg-shaped crown, an external lace (sometimes covered by tape) passing through eyelets attaches the webbing liner to the shell. A cloth-lined basil-leather to protect neck and sides of the face is sewn between the shell and the liner and fastened with a strap and buckle and sometimes additional pop-studs. An improved model has a somewhat flatter profile, while a raised rim with a squared inside groove protects the connecting lace from damage. First Canadian Army Routine Order of 27 July, 1941 ordered that all DR helmets be painted matt khaki green #3.

Efforts to procure DR helmets in Canada and the USA met limited success. An unknown number of an off-the-shelf, low-crowned fibre crash helmet with a padded rim was procured from the Canadian branch of Spalding Sporting Goods Ltd., in late 1942. There is no indication that these were used overseas, but they were still in use by RCASC and the Provost Corps motorcycle demonstration teams in the late 1950s. In late 1942, the Army Technical Development Board began a more formal process of developing a Canadian-made DR helmet, Project 1001-2. Requests for prototypes were sent to Spalding, Backstay Standard, and Dunlop, on 10 November, 1942. Spalding's helmet was rejected as the Diamond State fibre used in its helmet shell was too water absorbent. Backstay and Dunlop shells were more satisfactory, but preference was expressed for a shell made of phenol-formaldehyde impregnated sisal fibre submitted by the Sponge Rubber Co. in December. User trials of these helmets were very favourable.[17]

Figure 25 *A Canadian-made Helmet, Steel, DR Mk. I helmet painted semi-matt olive green. Note the wide chin strap and the slighly flared brim.*

Lucy Collection / Rowe Photo

Meanwhile the British had adopted a steel DR helmet. On 30 September, 1942, CMHQ advised National Defence Headquarters that the War Office had adopted three new steel helmets using a common shell the *Helmet DR, Steel, Mk. I* or *Helmet, Steel, No.3* for Despatch Riders, the *Helmet, Steel, No.4* for Armoured Fighting Vehicle crews and the *Helmet, Steel, No.5* for paratroopers. Production was to begin shortly, at which time samples would be available for Ottawa. On December 23, First Canadian Army advised CMHQ that the DR and AFV helmets would be adopted on the same scale as for the British Army. Ottawa was advised of this on December 26. On December 29, First Canadian Army asked the War Office for an allocation of 10,000 DR and 20,000 AFV helmets, as soon as they became available. Equipment Policy Letter No. 25 of 3 June, 1943 authorised their issue to all applicable personnel, and ordered the return of all non-steel crash helmets to depot.[18]

Again however, shortages continued to plague the issues of these helmets on the approved scale. In November, 1943, 2 Division had only 5% of the required number and 3 Division 25%. In any event, Canadian soldiers reportedly preferred the old pulp version and efforts were made by

Figure 26 *Post-war pictures of RCCS DRs in Canadian-made Helmet, Steel, Despatch Riders, Mk. I. The instructor seems to have a commercially acquired fibre crash helmet.*

NAC PA159849

CMHQ in late 1943 to get a dispensation from the War Office to continue its issue. CMHQ agreed, officially, to conform to British practice, but, on 22 January, 1944, permitted the continued use by Canadian troops of the pulp helmets until they were worn out, at which time they were to be replaced by the steel variety. Unlike their pulp predecessors the steel DR helmets really did bear a more than coincidental resemblance to German *Fallschirm* helmets, this, and the additional weight,[19] may account for some of their unpopularity.

At National Defence Headquarters' request CMHQ sent the specifications and samples of the steel DR helmets to Canada in April, 1943. Their arrival seems to have come as a surprise to Army Technical Development Board, but the design was enthusiastically received. The trials were very favourable, and in July, 1943 a Contract Demand for 14,000 DR Steel, Mk. I helmets was issued. On September 14, an OIC authorized the acquisition of the 10,000 steel shells from the Canadian Motorlamp Co. The shells were sole sourced on the grounds that, as helmet makers, the Canadian Motorlamp Co. had a substantially lower rejection rate than General Steel Wares or Aluminum Goods Ltd. For some reason it was not until August, 1944 that the contract for the liners was authorized. The contract was awarded to Backstay Standard. Sole-sourcing was justified on the seemingly incorrect grounds that Backstay Standard had the necessary experience in making helmet liners. At some point in 1943, the order was readjusted to 10,000 steel and 4,000 plastic helmets, and in late 1944, with the war in Europe ending, the order for liners to the steel version was cut back to 5,000 units.[20]

Figure 27 *A 1 Canadian Division officer wearing an early pattern British pulp DR helmet.*

NAC PA197540

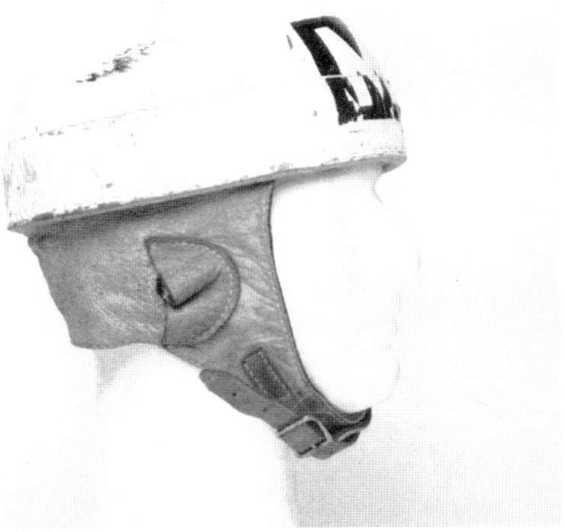

Figure 28 *A Canadian-issued fibre motorcycle helmet, made by Spalding in November, 1942. The helmet has a base colour of olive drab, but has been over-painted white, the white on black* MP *is painted over an earlier red on white* PROVOST. *This livery seems to date from the late 1950s.*

Lucy Collection / Rowe Photo

Known in Canadian service as the *Helmet, Crash, Motor Cyclists, Steel* or *A15550*, these helmets differ from British DR steel helmets in several minor ways; the base of the shell is a little larger and the rim is somewhat more flared. The shells have an inked C/Arrow stamp, and the marking C.L./C. 1944 and an L or S for the size impressed in the rear shell. The liners are stamped and are padded with sorbo rubber in lieu of felt, have a wider chin-strap than British made versions (18.5 compared to 12.5 mm), and are marked BS. The helmet shells cost $1.78 each to produce, (plus $10,233 for tooling) and the liners $6.75 each. The *Helmet, Crash, Motor Cyclists, Steel* remained in Canadian Forces stocks until 1980.[21]

The order of 4,000 fibre helmets (apparently to Backstay Standard - as their BS logo is found in the liners) was placed some time in 1944. Termed *Helmet, Crash, Motor Cyclists, Plastic* or *A15500*, it is similar in general form to the British *Helmet, Crash, Despatch Riders*, but has grommetted vents set in the centre front and sides of the shell while the liner band is attached by eight washered screw bolts set in the raised rim. The leather curtain, lined in cloth, has fixed ear guards and small gussets of rabbit fur at the temples, a concession, no doubt to Canada's climate. Additional production at the rate of some 6,000 per year was planned for 1945 and 1946.[22]

Since WWII Canadian military motorcyclists (and those using snow-mobiles) have used a variety of non-ballistic civilian or police style crash helmets. An effort was made in the 1970s to standardize on the *Crown 3000* Police Highway patrol type crash helmet, however the current policy is for off the shelf acquisition of commercial helmets at the unit level.

Figure 29 *Photograph taken in England early 1941. A soldier of the Essex Scottish wearing the formation patch painted on his helmet. The DRs helmet is probably from the batch acquired commercialy from Charles Owen & Co in 1940.*

NAC PA197138

Tank Crew

Initially the Canadian Armoured Corps in the UK was equipped with the British-made *Helmet, Crash, Royal Tank Regiment*, or the *Helmet Crash, Royal Armoured Corps*. The former was made of cloth-covered cork, with rubber vents in the sides. The latter, a war-time expedient introduced in late 1939 is made from 3 pieces of fibre rivetted together. Both helmets feature a "scrum" (a separate leather band for the head phones, attached by laces and buttons), and a crash pad in front. In the summer of 1940 procurement action was initiated in Canada for a tank helmet. A fibre model made by Spalding was tested by the Dress and Clothing Committee, but rejected in favour of a Canadian-made version of the *Helmet, Crash, RTR*. A requisition for 2,500 *Helmets, Crash, Tank Battalion* was signed by the ADM at National Defence on 30 July, 1940 and five firms were invited to tender, of which two: Safety Supply Co. of Toronto and Scully's of Montreal submitted bids. Safety Supply Co.'s bid, at $3.50 each, compared to Scully's $8.35, won easily and on 6 November, 1940, an OIC approving the contract was signed. Safety Supply Co.'s bid turned out to be wildly over optimistic, within a month it was complaining that it would cost $7.50 each to make the helmets, raising the cost of the contract from $8,750 to $18,750. As this was still lower than Scully's bid, the Privy Council approved the change on December 31, with the proviso that the costs be closely audited and if they proved lower than estimated by 5% that the contact be correspondingly revised downwards.[23]

As Canadian procurement was taking longer than anticipated (and because the London office of the Department of Munitions and Supply had mislaid its original requisition) the War Office was

Figure 30 *A Canadian Armoured Corps soldier entering the sponson of an M-3 Lee in England, circa 1941/42. He is wearing a British-made Helmet, Crash, RAC, without the scrum.*

NAC PA197538

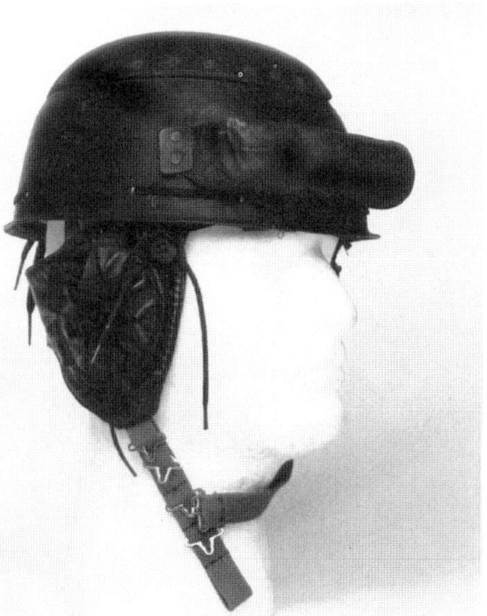

Figure 31 *A British-made fibre Helmet, Crash, Royal Armoured Corps, made by Helmets Inc., in 1940. It is black over the original matt olive drab, the scrum is black leather.*

Lucy Collection / Rowe Photo

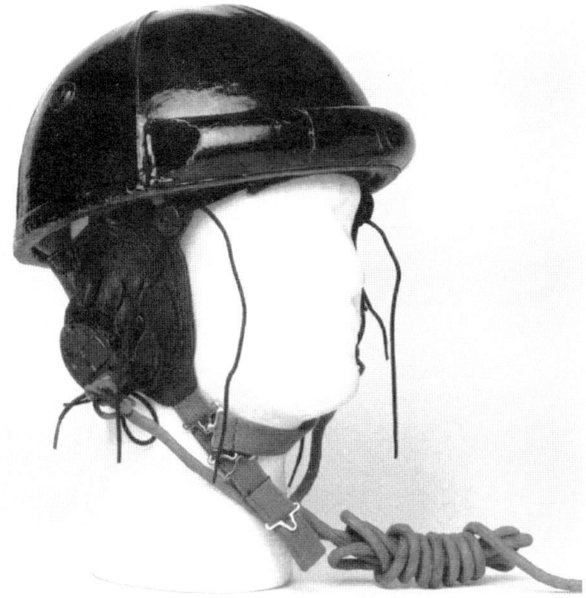

Figure 32 *A Canadian-made Helmet, Crash, Tank Battalion, a copy of the British Helmet, Crash, RTR. The canvas cover over the cork body is painted black.*
Lucy Collection / Rowe Photo

asked in late December, 1940 for a quotation on 2,500 *Helmets, Tank.* The quoted price of 16/10d each, with scrum and chin strap, was presumably for the cheaper fibre version, which Canadian troops are shown wearing in 1941.

Orders for *Helmets, Crash, Tank Battalion* from Safety Supply Co. were repeated; 3,000 in 1941, 8,000 in 1942 and a final 3,000 in February, 1943. In Canadian service these helmets received the stock number *A15600.* Production was at the rate of about 135 per month. They were still on Canadian Army inventory in 1949.[24]

CMHQ's Equipment Policy Letter 25 of 3 June, 1943 also confirmed the issue of *Helmet, RAC, Steel Mk. I (Helmet, Steel, No. 4*) to Canadian Armoured Corps units on the same scale as issued to the British Army. They were extensively worn in NW Europe and Italy. Their resemblance to the German Paratrooper helmet made them unpopular in areas where the *Fallschirmjaeger* operated. The helmet's Canadian designation was *IA19120.* As the Canadian Armoured Corps used large numbers of American M-3 and M-4 tanks, the US *M-6* helmet was also issued and is sometimes seen in photos. Its Canadian designation was: *Helmet, Crash, Tank, USA Pattern* or *A15650.* [25]

Efforts were also made to develop a winter tank helmet for Canadian use. Army Technical Development Board Project 1080 led to an experimental *Helmet, Tank Crew, Leather* in October, 1944, which was tested in Exercises Eskimo and Polar Bear the following month. The exercises revealed the need for additional padding and, in September 1945, the equipment received General Staff approval.

Figure 33 *Canadian Armoured Corps Captain wearing RAC Mk. I, Steel with Mk. II chin-strap, Italy 1943/44.*
NAC PA200059

***Figure 34** A fine shot of a CAC tank commander in Italy 1943/44 wearing the US M-6 crash helmet often worn by the crews of US made tanks.*

NAC PA197536

Procurement does not, however, seem to have been initiated. To date no information has emerged on what this helmet actually looked like[26] although specimens of an insulated, C/Arrow-marked helmet may be related to this project.

In 1948 the Department of National Defence issued a requirement for a tank helmet that would; offer protection from rifle and machine gun bullets and mortar fragments; come in 3 sizes; was as compact as possible; allowed use of head-sets, and; had a detachable liner made of non-conducting material.

The USA EX49 - EX51 tests were monitored *(see postwar section below)* with great interest but no satisfactory design emerged.[27] In the 1960s Canada adopted first the fibre-glass football helmet shaped *Gentex CVC* helmet (developed in the USA in the mid-1950s, after the efforts to develop a common all-arms helmet had been abandoned). This was found to be unsatisfactory (as was the similar British Racal AFV helmet, which was also considered) and was replaced by *Gentex DH-132*, made of thick neoprene rubber with a fibreglass shell, attached by Velcro, to give some ballistic protection. With the adoption of the Leopard tank in the early 1970s, Canada opted for the *Gentex DH-169* helmet, and later the slightly modified *DH-169A* crash helmet. The DH-169 differs from the standard DH-132 pattern by having a two piece skull and brow guard, to allow better access to weapons sights. It is equipped with an ambient noise pick-up which allows conversation while it is being worn but protects the user from sudden noise by cutting out at 85 db. The Canadian Forces are now adopting a new suite of communications equipment and as a consequence are looking for a new AFV helmet, which will probably be acquired off the shelf.[28]

***Figure 35** Method of stowing the US M-6 crash helmet inside an M-4.*

NAC PA197542

Figure 36 *A US-made M-6, which saw Canadian service. Made by Wilson Athletic Goods of Chicago. Made of fibre, the helmet is painted light olive drab with natural leather trim.*
Lucy Collection / Rowe Photo

Figure 37 *A British made Helmet, Steel, RAC, Mk. I. It is painted a sand textured dark charcoal grey.*
Lucy Collection / Rowe Photo

Figure 38 *A Gentex Helmet, Combat Vehicle Crewman, DH-169A AFV crew helmet. The fibreglass plates are coloured light olive green with black edging. This specimen was made in January, 1990. The DH-169A differs from the DH-169 by having a smaller microphone and other minor detail changes.*
Lucy Collection / Rowe Photo

Post War

Figure 39 *A 1960s vintage M-1 with UN decals, in white on blue. Note the clips which attach the chin-strap, a post-war feature.*

Lucy Collection / Rowe Photo

The Mk. III helmet and its successor the *Mk. IV* *were not adopted by the much reduced post-war Canadian Army. The design was considered too heavy and awkwardly shaped. In the short term, there were sufficient stocks of Mk. IIs on hand to meet peace-time requirements. Over the longer term, Canada planned, in the interests of standardization, to conform to USA equipment and weapons. The USA was conducting its own series of trials of fibre-glass and aluminium two-piece helmets, designed for common use by infantry, paratroops and AFV crews. Between 1949 and 1953 Canada and the UK carefully monitored the trials in the USA of the *EX 49 - EX 53* series of helmets, deferring further procurement decisions until the new USA standard was adopted.

The outbreak of the Korean war, to which Canada contributed a 3 battalion brigade to the Commonwealth Division, lead to a growing crisis in helmet supplies. In October, 1951 an inventory found only 40,000 Mk. II and 20,000 M-1s left in the depots. Inventories taken in 1954 found 11,778 assembled Mk. II and M-1 helmets in Ordnance stocks and 40,022 on issue to troops. An additional 72,000 were needed to meet M+3 months mobilization plans. While 86,938 bodies and 109,979 liners (including presumably some new ones produced in 1953 by Viceroy Manufacturing) were available in a disassembled condition, an estimated $100,000 was required to provide missing chin-straps, bolts, buffers etc. Many of the Mk. II (and other) helmets issued in this period are repainted with the same dark green paint used on Canadian Army vehicles. A substitute chin-strap made of light olive web with a brass slide buckle also seems to date from this time.

In the end, the USA found it impossible to find a single design which satisfied the needs of infantry, airborne and the armoured troops. While the experimental nylon and doron laminated fabric shells on the EX49/EX53 series were ballistically promising, the material used could not endure the other stresses of service use. The trials were eventually abandoned and the USA

* The Mk. IV was identical to the Mk. III except: the liner was attached by a press-the-dot fastener and the chin-strap lugs rivetted just above the rim, not ⅓ the way up as on the Mk. III. The idea was to use the helmet as a bucket in the Far East. The Mk. IV entered production in December 1944 and continued in production until late 1945. Some were issued to Canadian troops in NW Europe. Production resumed in 1952. In the 1960s a new liner was issued and the helmet retitled *Helmet, Steel, Mk.V*. This was never issued in Canada.

continued to use the M-1 until the 1980s. In 1960, a new, ballistically superior, nylon liner and improved suspension was adopted.[29]

Figure 40 *Canadian Airborne soldiers in the 1980s or early 90s wearing a modern M-1C with Canadian slide buckles and a US reversible leaf pattern camouflage cover.*

Photo courtesy Ed Storey, 208-36

In 1952 the British re-opened the Mk. IV production line and Canada may have ordered a limited quantity of them. Reportedly, some were worn in Korea and others issued to Militia units (the Ontario Regiment wore them on parade into the 1980s.) Some interest was shown in further British helmet developments. In February 1957, Canada requested two examples for trials of a new liner being developed by the UK for the Mk. IV helmet (the prototype of the British Mk.5).[30]

Figure 41 *A Van Doo in the 1960s wearing a war-time M-1 with a regimental decal.*

Photo courtesy Ed Storey

With the US trials making so little progress, the Canadian Chief of Staff directed the Quartermaster General, on 25 May, 1953, to investigate the development of a helmet made from *Canadian Flexible Armour*. This was a ballistic material comprising layers of ribbed nylon weave and an experimental ballistic fabric called "Fortisan", or *X-36*, which the Canadian Army was using to develop an armoured vest. Canadian Flexible Armour (which, despite its name, was actually produced in the USA) showed good ballistic characteristics, and helmets made from it would, it was assumed, offer much better protection than that provided by the Mk. II or even the M-1 helmet. One problem, of course, was that Canadian Flexible Armour was just that, flexible. Military helmets need to be rigid to protect against trauma. The first mock-ups assembled by the Directorate of Inter-Service Development (DID), consisted of resinated canvas inner and outer shells (to give the needed rigidity and to make the whole water-proof) with three layers of corded nylon and six of X-36 sealed in-between. These mock-ups

Figure 42 *A Canadian corporal late 1980s early 1990s in a late model M-1 with woodland camouflage cover and Type-1 chinstrap.*
Photo courtesy Ed Storey

performed quite well in informal ballistic tests (a v50 of 1399 f/s). On 6 November, 1953, authority was obtained from the Army Equipment Policy Committee to expend $3,000 to refine the design and produce prototype examples for ballistics and engineering trials.

In June, 1954, an initial series of eight examples (four with two layers and four with three layers of Canadian Flexible Armour) was made up under supervision of a DID officer by Mine Safety Appliances of Pittsburgh, Pennsylvania (a company that produced many of the M-1 liners as well as specialized safety helmets). Quotations for producing an additional 125 examples were also sought. Thirty of the helmets were subjected to ballistics tests at some time between August 1954 and August 1955. These indicated v50 results between 1387 and 1510 f/s (about 410-450 m/s). Money was appropriated for another 100 helmets designated the *X-55-5*. The X-55 resembled the US M-1 in form, and was made of 3 layers of ribbed nylon weave and 7 of X-36, bonded with phenolic resin. It weighed almost 3 lbs (47.5 oz) i.e. about 1.4 kg. of which 86% was the fabric, 14% resin. It was painted olive drab. No firm information on the X-55's suspension and chin-strap arrangements is provided, both M-1 and Mk. II types seem to have been were tested. A 1956 report refers to its using a fibre glass ring to hold the suspension. There was also a lightweight version, the *X-55b*

Figure 43 *A Canadian M-1C from the 1970s, with the slide buckles to fasten the inner shell's chin-strap. It is covered with a leaf pattern camouflage cover and a scrimmed nylon net.*
Lucy Collection / Rowe Photo

weighing 14.2 oz (about 400grams), described as being similar to, but smaller and denser than, an experimental nylon liner being tested by the USA at that time - the *T-54*.

The X-55 was subjected to ballistic trials at the US Army's Watertown Arsenal in October 1955. These showed a v50 of 1203 f/s on the crown and 696 f/s in the brim area (400 and 230 m/s respectively). The former figure compared favourably with the old Mk. II (883 f/s or 270 m/s) and the M-1 with steel pot and standard fibre liner, however it was inferior to the v50 of 1331 f/s (440 m/s) displayed by an M-1 equipped with a *T-55-2* nylon liner. This latter combination was to become the US Army standard by the early 1960s. The X-55b was estimated to have a v50 of about 1100 f/s (330 m/s), but does not seem to have been actually tested. An aluminium shell seems to have been made to go with it. In 1956 an additional 100 helmets (now termed *X-56*) and 70 harnesses were made up and tested by the US firm Gentex (General Textiles) of Carbondale, Pennsylvania. In the summer of 1956, 50 of these helmets were issued to the Canadian School of Infantry at Camp Borden for user trials. Others were to be used for engineering trials. The outcome of these trials is not known. Work on the Canadian Flexible Armour helmet was terminated around 1958, due to technical difficulties and problems with property rights to the design. The M-1 was officially adopted as the Canadian standard on 20 September, 1960, but it had been slowly supplanting the Mk. II throughout the 1950s. M-1s were, for example, used on manoeuvres by troops playing the opposing forces. These were usually marked by a coloured triangle or some other geometric symbol. M-1s in this period and the early 1960s were often worn with a regimental decal on the front or at the side of the shell. For field exercises the war-time bi-coloured net, or US-made nets or camouflaged covers (such as the USMC snipers veil cover) were often worn.[31]

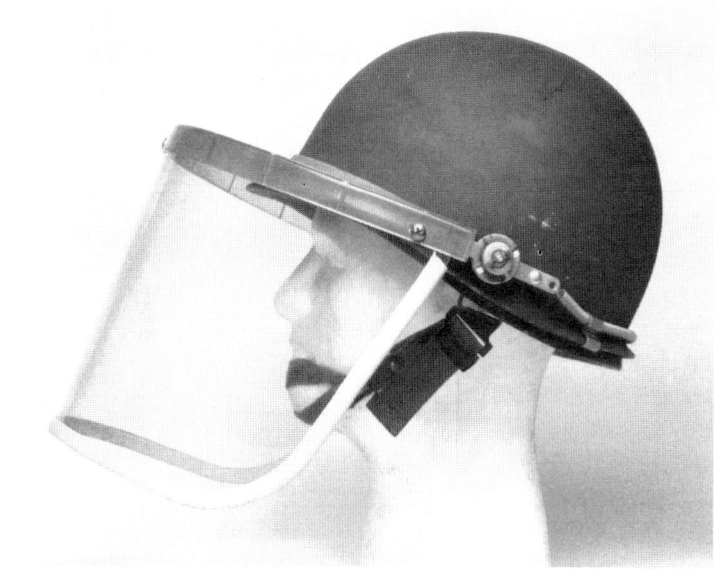

Figure 44 *A 1970s vintage M-1 with a riot shield. The helmet is painted light olive green, note the Type 1 chin-strap*

Lucy Collection / Rowe Photo

Under the North American Defence Production Sharing Agreement, much of Canada's defence production is shared with the USA. Among the goods manufactured under this arrangement were steel outer shells for the M-1, produced by R.J. Stampings of Smith Falls, Ontario. The tooling for these shells was identical in every respect to post-1960s USA-made helmets. Much of R.J. Stamping's production was, in fact, shipped to the USA and their helmets bear no particular markings to distinguish them from shells made in the USA. The liners for Canadian Forces' M-1 helmets were procured from the USA, with some minor adaptations to the fittings to optimize them for Canadian use (see below with reference to paratrooper helmets). Canadian troops have been issued wartime and post M-1 helmets and liners of almost every conceivable variety and combination. Those acquired after the 1960s use the nylon liner and crossed web suspension that was the final outcome of the US T-55 development. After 1970, the *Type-1* chin-strap, with a snap fastener and a web chin cup, which clips on to the outer shell's chin-strap brackets came into service in Canada as well as the USA. From the same period Canadian helmets began to sport US-made leaf or woodland pattern camouflage covers, sometimes with scrimmed nylon nets (recycled laundry nets). The old bi-coloured net remained on issue until 1975. The French style scalloped reversible plastic "*filet de camouflage en synthetique*" was also used. A white nylon cover is often worn in winter conditions.[32]

Figure 45 *Canadian Airborne soldiers in North America wearing the inner liners of the US M-1C, note the leather chin-cups.*

NAC PA197532

Airborne Troops

Canada raised an Airborne battalion in WWII which served with the British 6th Airborne Division. Equipment Policy Letter 25 of 3 June, 1943 confirmed that the British *Helmet, Steel, Airborne Troops, Mk. I* (*Helmet, Steel, No.5*) would be issued to Canadian paratroopers. These helmets went through a number of changes. The first issues had a fibre rim, relaced by late 1942 by one of stainless steel. The AT Mk. I has a four-point leather suspension screwed to the shell. By D-Day some Canadian paratroopers had received the *Helmet, Steel, Airborne Troops, Mk. II*, with a three point web suspension. Canadian designations for these helmets were: *Helmet, Paratroop, P-type, British* or *IA17150*, and *Helmet, Steel, Parachutists* or *IA19100*. For training purposes in the UK, the canvas and sorbo rubber *Airborne Troops Training Helmet, Standard Pattern* was also issued. In addition the Canadian Airborne Training Centre in Rivers, Manitoba also used, in addition to their UK equivalents, the US Riddle plastic training helmet and the steel *M-1C* with the former being rejected as too fragile*.[33]

* The M1-C resembles the M-1 in every respect, except the inner liner has a pop-stud on each side to snap to corresponding studs at the end of the outer shell's chin-strap, and an A shaped web straps rivetted to each side of the liner, with brass wire buckles which attached a leather or web chin-cup.

Figure 46 *The Riddell, see-through, football-style helmet. This was the pattern initially used by Canadian trainees at Fort Benning, Georgia.*
MilArt Photo Archive

Immediately after the war Canada's airborne forces were reduced to a single company drawn from the Princess Patricia's Canadian Light Infantry (PPCLI). By the end of the decade, however, Canadian defence policy had changed to meet the possible threat posed to Canada's vast Arctic territory by a potential Soviet aerial descent. To counter this perceived threat, a much expanded airborne capacity in the form of a rapid reaction Mobile Strike Force (MSF). The MSF was envisaged as a force able to operate anywhere in Canada, including the Arctic, but its formation required the expansion of Canada's airborne cadre from company to battalion and then to nominally brigade strength. Each of Canada's three Regular Force infantry regiments: the PPCLI, the Royal Canadian Regiment (RCR) and the Royal 22e Regiment ("Van Doos") contributed a battalion, intended to be jump or glider capable, to the MSF. This put great stress on the supply of a variety of airborne equipment from transport aircraft to helmets.

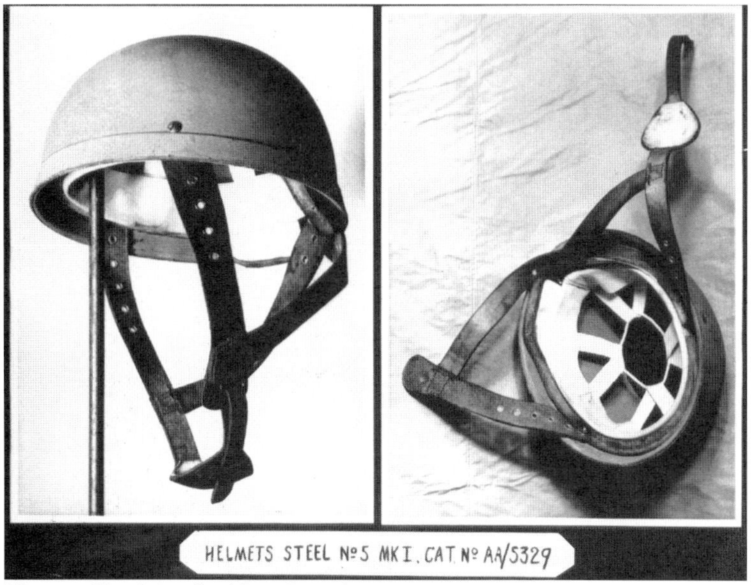

Figure 47 *March, 1944 British Army stores photo of an early fibre-rimmed Helmet, Steel, Airborne Troops. These were manufactured prior to the adoption of a stainless steel rim in September, 1942. They were painted a pale grey green colour. Some were issued to early members of the Canadian Airborne Battalion.*
MilArt Photo Archive

In 1951-2 the MSF required 3,000 Paratroop helmets. The receipt of 900 helmets from the UK in 1948 had provided sufficient stocks to equip the PPCLI battalion (which had 819 "P" type helmets on hand on August 3, 1951), but there were not enough to equip the other battalions, or the Joint Airborne Training Centre (JATC) at Rivers, Manitoba. An order for 400 new helmets was placed with the British War Office in June, 1951, but the War Office replied that airborne helmet

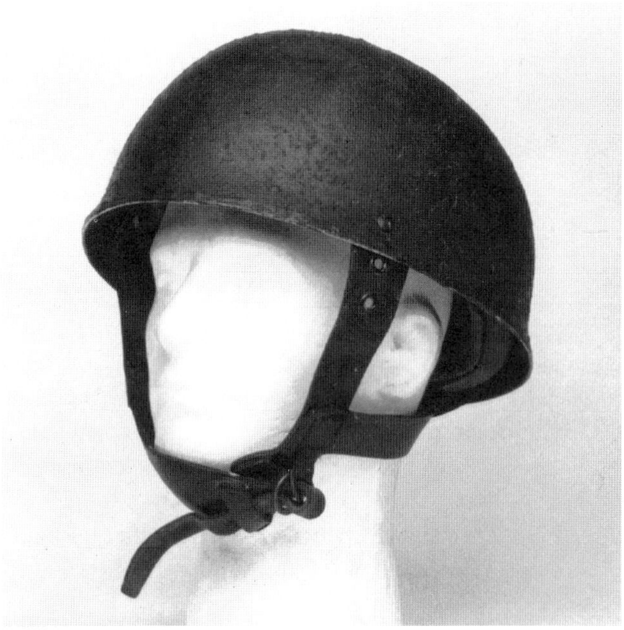

Figure 48 *Helmet, Steel, Airborne Troops Mk. I converted from a Canadian-made DR Steel Mk. I. Painted sand textured dark olive drab, this seems to be a post war conversion designed to address the shortage of para helmets. Note the helmet is worn reversed, with the DR's brown pad becoming the neck support for the para helmet. It retains the original DR helmet liner, on a steel band rivetted to the shell.*

Lucy Collection / Rowe Photo

In light of the MSF's mission to defend the Canadian North, National Defence was also interested in an airborne helmet suitable for Arctic use. Various expedients were tried, such as the British Sorbo training helmet, five examples of which were evaluated, along with four ice-hockey helmets (a truly Canadian solution) which were acquired for trials in August, 1951 at $31 each, and designated *Helmets, Parachutists, Experimental.*

With the collapse of the USA EX-51 project, the UK re-opened its helmet production lines. On 30 January, 1953, 4,000 new-production *Helmets, Steel, Airborne Troops, Mk. II* were ordered from the UK and were delivered by the end of the year. By this time the supply situation was so desperate that another production had ceased at the end of the war. Pending the outcome of USA efforts to design a helmet suitable for airborne, infantry and armoured use, the British had no plans for procuring new helmets for its own use. The Canadian order was too small to justify reopening the line.

To meet this deficiency a number of Canadian-made DR Steel Mk. I helmets were converted to airborne use and issued to the Van Doos and RCRs. The conversion consisted of inserting a top pad, cutting off the curtain and chin-strap and boring 4 holes in the rim to take the leather Airborne Mk. I harness. The helmets were worn reversed so the brow pad on the DR helmet became a neck support on the Airborne helmet. Reports from the units indicated that these conversions were considered very unsatisfactory.

Figure 49 *A Canadian para in 1944 with Sten Mk.V and Helmet, Steel, Airborne Troops Mk.II covered by a heavily scrimmed woven net.*

NAC PA179157

Figure 50 *This Helmet, Steel, Airborne Troops, Mk. II started life as a Mk. I and the holes for the leather harness can still be seen. It features a Canadian made liner. The helmet is painted white with a red, gold and white PPCLI decal.*
Lucy Collection / Rowe Photo

Figure 51 *A British-made Helmet, Steel, Airborne Troops, Mk. II (BMB 1952) with its original liner. Overpainted in the dark green paint used on post-war Canadian Army vehicles it has the Van Doos decal on the left side.*
Lucy Collection / Rowe Photo

Figure 52 *This DR helmet has been converted to a Helmet, Steel, Airborne Troops, Mk. II, with a proper (Canadian made) para liner and web harness. Note the redundant holes for the DR liner and the Mk. I type leather harness. The helmet is painted matt dark olive drab, with a white rim and lightening flash.*
Lucy Collection / Rowe Photo

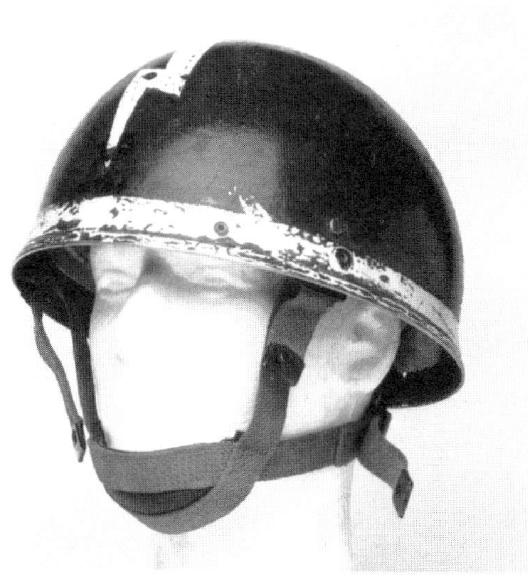

Figure 53 Part of a "stick" from the R22eR prior to emplaning. The three paratropers in overalls have had their names taped to the front of their helmets. Quebec, circa 1953.
Photo courtesy Maj. R.L. Fournier, CD

100 DR steel helmets had to be loaned to the JATC to meet its urgent needs.[34] Even this replenishment cannot have met all requirements for paratroop helmets. One encounters from this period old helmets which have been refurbished to Mk. II standard, as well as post-war Mk. IIs that have been locally overhauled. These can be identified by their substitute sweat bands and crumbly yellow foam-rubber padding. The British Airborne helmet remained in service until 1970, when the latest version of the US M-1C replaced them. By 1978 a considerable number of these M-1Cs were modified for Canadian use by replacing the brass wire claw buckle on the inner shell's chin strap with a slide buckle and cutting off the eyelet ends of the chin support. This helmet remains in use with airborne-capable companies of the Light Infantry Battalions which have replaced the Canadian Airborne Regiment.

Experimental Helmets Tested by Canada

In terms of its longevity and widespread production, the M-1 helmet has been an unrivalled success, but after almost 50 years in service it has been overtaken by advances in materials and design and by changing requirements. From the late 1970s a number of Western countries began to develop new helmets made of composite materials or ballistic plastics. The best known of these is, of course, the US *Personal Armour System, Ground Troops* model (*PASGT*) colloquially known as the "Fritz" from to resemblance to the old German *Stahlhelm*. In the 1990s, Fritz-type helmets became the type which other countries, including Canada, imitated in their search for new helmet

Figure 54 A post war Helmet, Steel, Airborne Troops, Mk.II painted vehicle green with a white red painted insignia for airborne units of the RCE.
Storey Collection / Photo

designs. In 1984, Canada began a series of trials to identify or develop a modern combat helmet suitable for ground troop and airborne use. Initially, a wide variety of existing steel and composite helmets were tested for suitability taking into consideration; ease of manufacture; stability and wearability; compatibility with existing equipment, and; ballistic resistance, in order to determine the features to be incorporated in a new Canadian helmet. Materials considered included, as well as steel, various composites such as; Aramid nylon (ie Kevlar or Teuron); polyethelene (Spectra, Dyneema or Corlon); fibreglass and metal/composite mixes. These composites consist of layers of woven fabrics bonded with resin and moulded into shape. This allows more scope for designing a ballistically desirable shape than pressed steel. Conceptual models were then produced in very limited quantities to determine the best possible shape and material. An example was the *CLF-1ss* a composite helmet with a styrofoam liner and plastic straps tested in 1987.

Figure 55& 56 below *Barrday "Prototype P2 Soldiers Helmet" used in trials in 1991. It is painted sand mix light olive green. Note the very elaborate harness, the M-1 type liner (with BARRDAY INC. stamped on the crown band) bolted to the shell and the styrofoam trauma pad inserted in the top. Note also the thickness of the spectra composite shell. Some Barrday prototypes have a liner made entirely of squares of thick grey styrofoam.*

Lucy Collection / Rowe Photo

The process of final development was put out to tender. The winning bid went to the Barrday Co. of Cambridge, Ontario. Barrday was then a subsidiary of Allied Signal, which had developed the polyethylene-based ballistic composite Spectra. Barrday acquired the necessary pressing equipment (although the first batches were pressed in the USA at the facilities of the Devil's Lake Sioux - a major PASGT manufacturer) and between 1988 and 1990 produced a series of helmets for ballistic, engineering and user trials. Shells were fabricated from a variety of woven or unidirectional fibres. Versions were tested with both styrofoam liners, which provide better trauma protection (at the expense of good ventilation) and more conventional web suspensions. Helmets in this *P* series had flatter sides and a higher back than the PASGT which they superficially resemble. They

weigh between 1.2 - 1.6 kg.

User trials included limited issue to some of the Canadian peacekeeping contingents, including the Royal Canadian Dragoons in Somalia and the RCMP in Former Yugoslavia. Problems reportedly encountered with this design included; the helmet's rear rim not clearing the collar of with the new Canadian armoured vest when the soldier was firing in a prone position; and "delamination" of the shells under field conditions. The contract was cancelled in 1992 after Allied Signal sold Barrday. Barrday, however, still offers helmets of this design for police and commercial export markets.

In the meantime, to meet the needs of Canadian peacekeepers, successive off-the-shelf purchases of PASGT helmets were made from the USA. Satisfaction with the PASGT has been less than total. Its two-point harness lacks stability, there is (shades of 1943!) insufficient room to permit the wearing of the current pattern of headphones, and painting the helmet (i.e. UN blue) damages the resin matrix and leads to delamination. A new helmet design was therefore undertaken. Prototype testing with 2 Canadian Mechanized Brigade Group personnel at Petawawa was completed in May 1996 and the winning bid was made by the French company Gallet, which produces the TC-3 helmet for the French Armed Forces. The TC-3 marries the PASGT shell with a French liner and harness (similar to that used on the steel Mle-78 series). The TC-3 has been adapted for Canadian use by reducing the front peak to accomodate the C79 optical sight used on the C7 rifle and widening the flare at the sides so that it can be used with Canadian headsets, it is also compatible with the

***Figure** 57 Modelling proposed new equipment a Canadian soldier in the 1990s in a US-made PASGT.*
Photo courtesy Ed Storey, C43-2

***Figure 58** The new Model CG634 helmet, shown here with camouflage cover and optional headset package. Note the head-set retaining strap which goes over the top of the helmet.*
Photo courtesy Department of National Defence

C4 gas-mask. This new helmet, designated the *CG 634*, is made of Aramid (Kevlar) and has a minimum v50 of 634 m/s (compared to the PASGT's 610m/s). National Defence is acquiring 60,000 CG634 helmets which will be manufactured at a plant established by Gallet at St. Ramuald, Quebec. Issues are scheduled to begin in the spring of 1997, to be completed in 1998.[35]

Figure 59 *Sixty-five years of Canadian helmets. A green painted rimless Brodie helmet with the flash of 4 Canadian Machine Gun Batallion (a red arrow on a green square), a 1942 vintage M-1 issued to an officer from British Columbia and the Barrday P2.*

Lucy Collection / Rowe Photo

Bibliography

Primary documents

Files held in Record Groups (RG) 2 (Privy Council Office), 9 (Militia & Defence) and 24 (National Defence) by the National Archives of Canada, as well as files held by the Department of National Defence, Directorate of History & Heritage (DHH).

Published works

Cotton, Marcus, *British Steel Helmets, 1915-1918*, in two parts, *Militaria* Magazine, Collectors Press Ltd., London, UK

— *WW2 British Steel Helmets, Mk. I* and Mk. II*, *Militaria* Magazine, Collectors Press Ltd., London, UK

— *WW2 British Steel Helmets, Mk. III and Mk. IV, 1945*, *Militaria* Magazine, Collectors Press Ltd., London, UK

Powers, D.B., *Casques du Combat du Monde, tome I*, Lavauzelle, Paris, 1984.

Reynosa, Paul, *The M-1 Helmet*, Schiffer, Atglen, PA, 1996.

Notes

1. NAC, RG 24, Vol. 1871, SF 8.
2. *British Steel . . . Op cit.* Pt 1 *Militaria # 4,* April 1994, pt 2, *Militaria* # 6, July, 1994.
3. *Casques du . . . Op cit.*
4. NAC, RG 9 III C1, Vol. 3833, folder 17, file 1-3; RG 24, Vol.1871, GAQ 11-67.
5. NAC, RG28, Series A, Vol.12. Also *WW2 British Steel Helmets, Mk. I* and Mk. II, Militaria* # 9, October, 1994.
6. NAC, RG 2, microfilms 5124 ff. OICs 2206, 2207, 2208; RG28, Vol.28 file 1-1-209.
7. Orders-in-Council (OICs) *1941*, 1409, 1769, 3795, 4277, 4700, 4776, 4822 and 9879; *1942*, 708, 709, 2034, 2196, 2454, 2515, 2790, 3681, 3786, 4277, 4512, 4514, 4870, 4884, 4885, 4886, 4962, 5277, 7499, 7529, 8073, 9072, 10069; *1943*, 6119; *1944*, 4848; *1945*, 292; RG28, Vol.28, file 1-1-209.
8. NAC, RG 28, Series A, Vol.12; Vol. 34 file 1-1-14; RG 24, Vol. 5739, file S/8723-3-5.
9. NAC, RG 2, OIC 1944, 4848; RG 2, Vol. 10,047, file 13/Clothing/4.
10. *British Steel Helmets Mk. III and IV, 1940-45, Militaria # 14* April, 1995
11. NAC, RG 24, Vol. 10052, 13/Cost/2; file 55-1640P (on microfilm reel 5786). (DHH) 112.3 H1.009 (D284).
12. NAC, RG 24 Vol. 12,232, 1/Helmets/1; file 55-1640P (on microfilm reel 5786).
13. *Ibid.*
14. NAC, RG 24, Vol.5735, file 8723-3-5.
15. *The M-1 Helmet, Op cit.*
16. NAC, RG 24, Vol. 5739, file 8723-3-5; Vol. 19,407, file 2280-1.
17. NAC, RG 24, Vol. 19830, file 8937-5-2. Personal communications Ed Storey, Ken Joyce.
18. NAC, RG24, Vol.5741, file 8723-3-5; Vol. 19,830, file 8937-5-2-1 part 1. DND Directorate of Public Relations.
19. Personal Communications NDHQ/ DSSPM, Ed Storey.
20. NAC, RG 24, Vol. 13243, 13/Cloth/4ff.
21. *The Winged Wheelpatch*, pp 125, 129; DHH, 115.4101 3 (D88).

22 NAC, RG 24, Vol. 12,232, 1/Helmets/1.

23 *Ibid.*

24 NAC, RG 2, OICs *1943*, 7210, *1944*, 4859; *1945*, 4745.

25. NAC; RG 2, OIC, *1945*, 4745.

26 NAC, RG 28 Series A, Vol.12; DHH 115.41013b (D88), Report on Project 1001-2.

27. NAC, RG 2, OICs *1940*, 6247 and 7711
28. NAC, RG 2, OICs *1941*, 7383; 1942, 6717 and 11525; *1943*, 9294. RG 24, Vol. 13243, 13/Cloth/4ff.
29. NAC, RG 24, Vol. 5739, file 8723-3-5; microfilm reel C-5786 file 55-1640P.
30. DHH, 115.41013 (D99).
31. NAC, RG 24, Vol. 5739, fil 8723-3-5.
32. Personal communication NDHQ DSSPM.
33. NAC, RG 24, microfilm reel 5786, file 55-1640P.
34. NAC, RG 24, Vols. 5739-5741, file 8723-3-5.
35. Personal communication NDHQ DSSPM.